How Do Psychoactive Drugs Affect the Body?

Understanding How Drugs Affect the Synapse and Body Organs

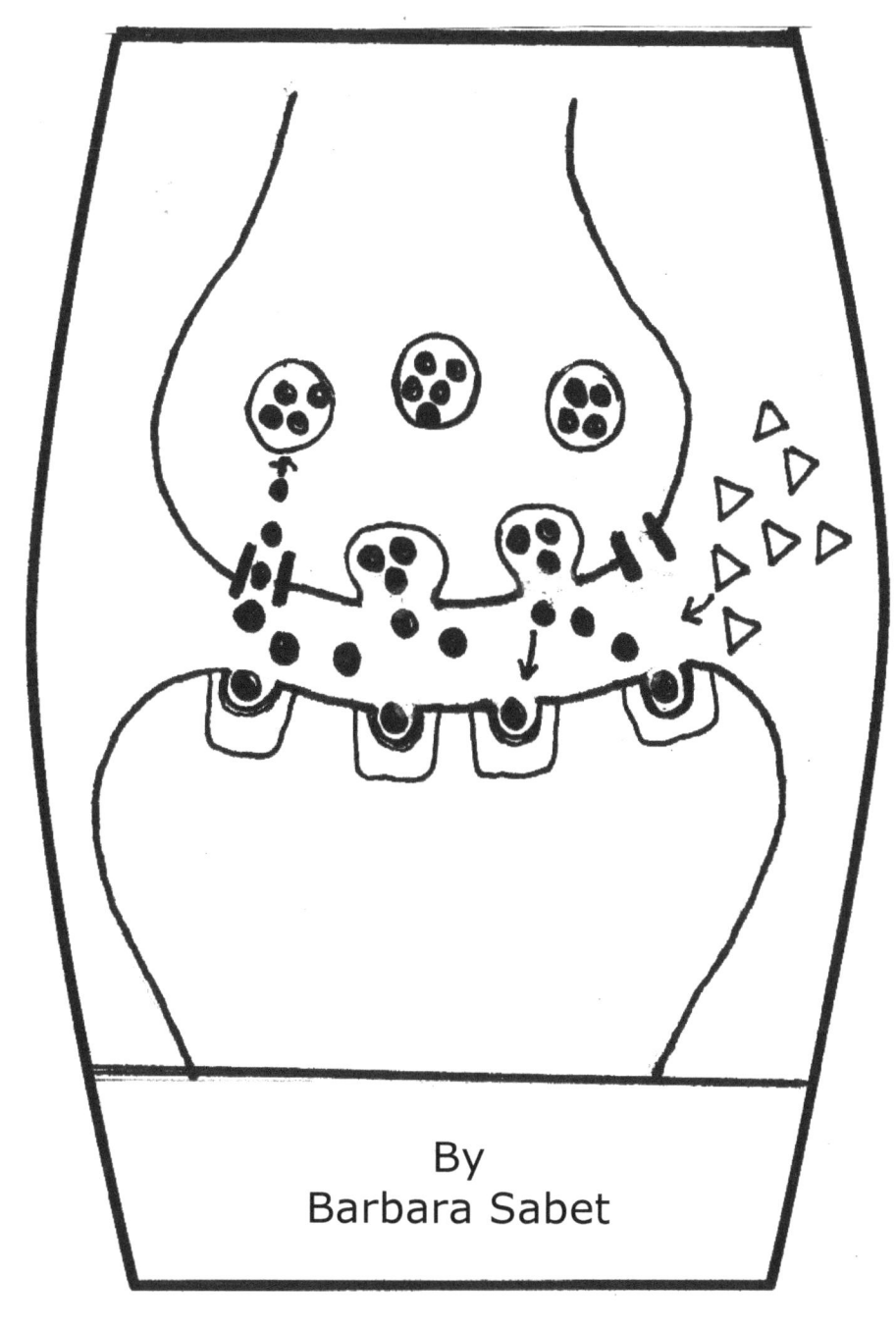

By
Barbara Sabet

Table of Contents

Introduction

When learning about the effects of psychoactive drugs on the brain and body, it is helpful to have an idea of how they affect the synapse. The synapse is the space between neurons where neurotransmitters (chemical messengers) enter and move across to the next neuron in the pathway.

You will see that drugs can do various things to neuro-transmitters, such as mimicking them, stimulating them, inhibiting them, releasing them, blocking them, etc. We are concentrating on psychoactive drugs in this book, but other prescription drugs affect the neuron and synapse in similar ways.

In this book, psychoactive drugs that are abused by us have been separated into their respective groups which are depressants, stimulants, opioids, hallucinogens and cannabis. Under each of these groups, one or more drugs have been chosen as examples, depending on their prevalence, as subjects of a single-paged worksheet. The worksheet shows how major body organs are affected by the drug and shows in detail how the synapse is affected. In filling-out the worksheet, the student should use the information found in the introduction to the drug groups, books, internet sources and teacher lessons.

Sample answer sheets are included at the back of the book for the teacher to use as a guide; or if this book is purchased for student independent study, the student can use them to help them fill-in the worksheets. The student should feel free to add to each worksheet as they find more information.

How the Synapse Between Neurons Works

The synapse is a fundamental part of the neural pathway because it is the way in which a signal is transferred from one nerve to another. Through it's neurotransmitters (NTs), the synapse regulates decision-making in terms of exciting or inhibiting the post-synaptic neuron. NTs are chemical messengers that move across the synapse from the pre-synaptic knob to bind with receptors on the post-synaptic membrane of the next neuron. They are stored in vesicles within the pre-synaptic knob. NTs can be *excitatory* or *inhibitory*. This means that they can either excite the post-synaptic neuron by contributing to depolarization and propagation of the action potential or they can inhibit the post-synaptic neuron by hyperpolarization and prevention of the action potential.

Look at the generalized diagram of the synapse below using EXCITATORY NTs. As you read the following steps to how a nerve impulse is propagated, follow along on the diagram. Dopamine and acetylcholine are examples of excitatory NTs.

STEPS in Propagation of a Nerve Impulse in an EXCITATORY Pathway:
1. Action potential moves down the axon
2. Depolarization causes the calcium channels to open and calcium ions enter.
3. Ca+ ions cause the vesicles to fuse with the pre-synaptic membrane.
4. The neurotransmitters (NTs) are released via exocytosis into the synapse and bind with the receptors on the post-synaptic membrane.
5. Na+ ions flow into the post-synaptic membrane
6. The membrane potential depolarizes and the action potential is propagated.
7. NTs are pumped back into the knob through re-uptake channels

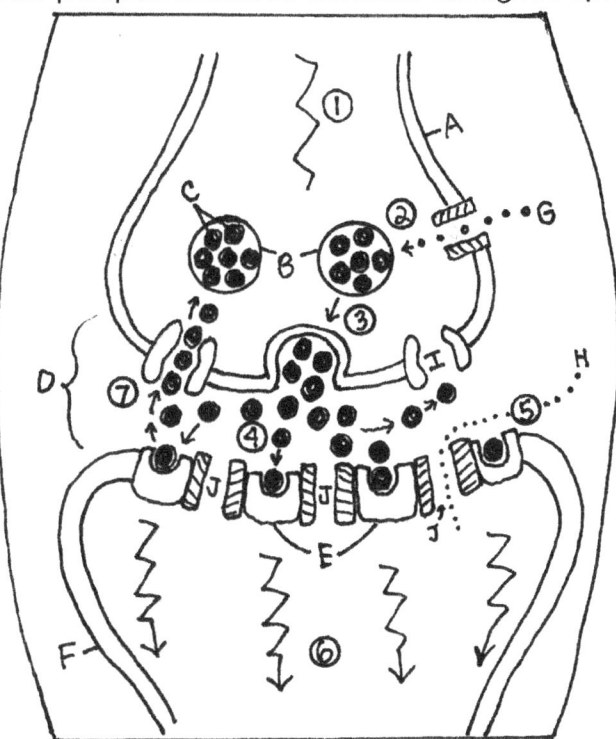

Key:
A. pre-synaptic knob of neuron
B. vesicles
C. neurotransmitters
D. synapse
E. neurotransmitter receptors
F. post-synaptic neuron
G. Ca+ ions
H. Na+ ions
I. re-uptake channel
J. ion channel

Look at the diagram below to see how an INHIBITORY neurotransmitter works. Look to see how it is different from the excitatory pathway on the previous page. GABA (Gamma-Aminobutyric Acid) is an important inhibitory NT. It opens Cl- ion channels on the post-synaptic membrane, which hyperpolarizes the post-synaptic neuron, and thereby reduces the action potential. In this way, it regulates the nervous process and calms or depresses the system. It prevents neurons from overfiring. Sometimes, doctors give a patient GABA in pill form to help with anxiety and stress-related disorders.

STEPS in the INHIBITORY Pathway:
1. Action potential moves down the axon.
2. Depolarization causes the calcium channels to open and calcium ions to enter.
3. Ca+ ions cause the vesicles to fuse with the pre-synaptic membrane.
4. The neurotransmitters (NTs), like GABA, are released via exocytosis into the synapse and bind with special receptors on the post-synaptic membrane.
5. GABA (or other inhibitor NTs) opens chloride (Cl-) ion channels on the post-synaptic membrane.
6. Cl- rushes in, and causes hyperpolarizing (rather than depolarizing) of the post-synaptic neuron, **reducing** action potentials.

Close-up of GABA-gated Cl- Channel
GABA actually attaches to Cl- channel

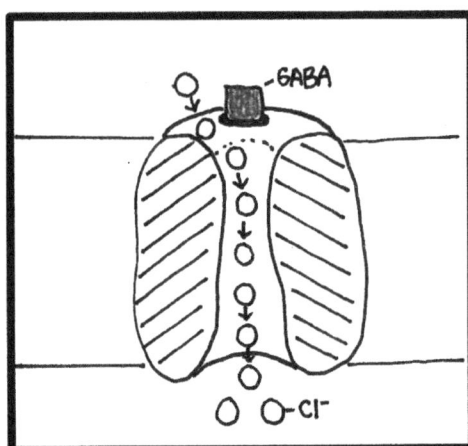

No action potentials due to hyperpolarization

Key:
A. pre-synaptic knob of neuron
B. vesicles
C. neurotransmitters (GABA)
D. synapse
E. GABA receptors
F. post-synaptic neuron
G. Ca+ ions
H. Cl- ions
I. re-uptake channel
J. ion channel

Mechanism of Drug Effects on the Synapse

Psychoactive drugs can affect the synapse in many ways and are either excitatory or inhibitory in their actions. They affect the brain and personality by either increasing or decreasing post-synaptic transmission through varied actions. They act on the synapse by inhibiting (antagonists) or exciting (agonists) neurotransmitter systems.

Look at the diagram of the generic synapse below. The 4 main ways that drugs affect this sytem are numbered in the corresponding area on the diagram. Drugs that act in each way are mentioned as examples too.

Location of Drug Interaction with Neurotransmitters (NT)

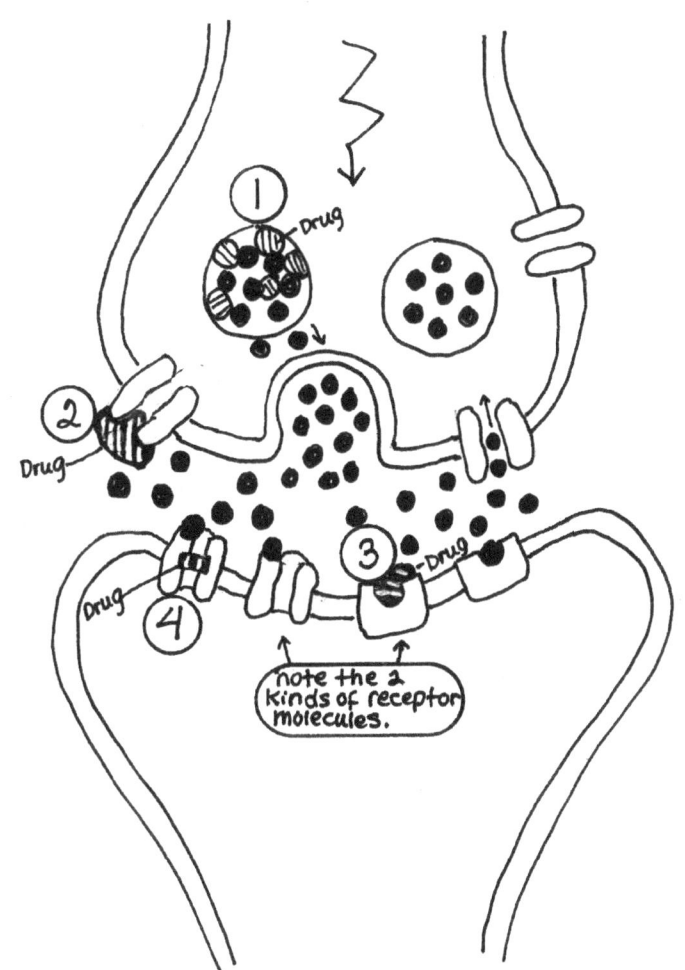

location #1: Drug Stimulates or inhibits the release of a NT. i.e. **methamphetamines** and **amphetamines** invade vesicles, push out extra *dopamine* into the synapse. The drugs also stimulate vesicle fusion with the membrane = EXCITATORY. dopamine causes pleasure

Location #2: Drug blocks the re-uptake of a NT. i.e. **cocaine** blocks *dopamine* re-uptake so that it remains in the synapse exciting the post-synaptic nerve = EXCITATORY

Location #3: Drug is a receptor blocker (it mimics the NT) or is a receptor stimulator. i.e. **Nicotine** mimics *Acetylcholine* and excites the neuron to fire = EXCITATORY. i.e. **Heroin** mimics and inhibits *Substance P*, a NT that causes pain = INHIBITORY

Location #4: Drug stimulates or inhibits ion channels in the post-synaptic membrane. i.e. **PCP** blocks the channel associated with *glutamate* (an excitatory NT) = INHIBITORY. The lack of *glutamate* causes dizziness, hallucinations, confusion, etc.

Worksheets on Psychoactive Drugs and How They Affect the Body

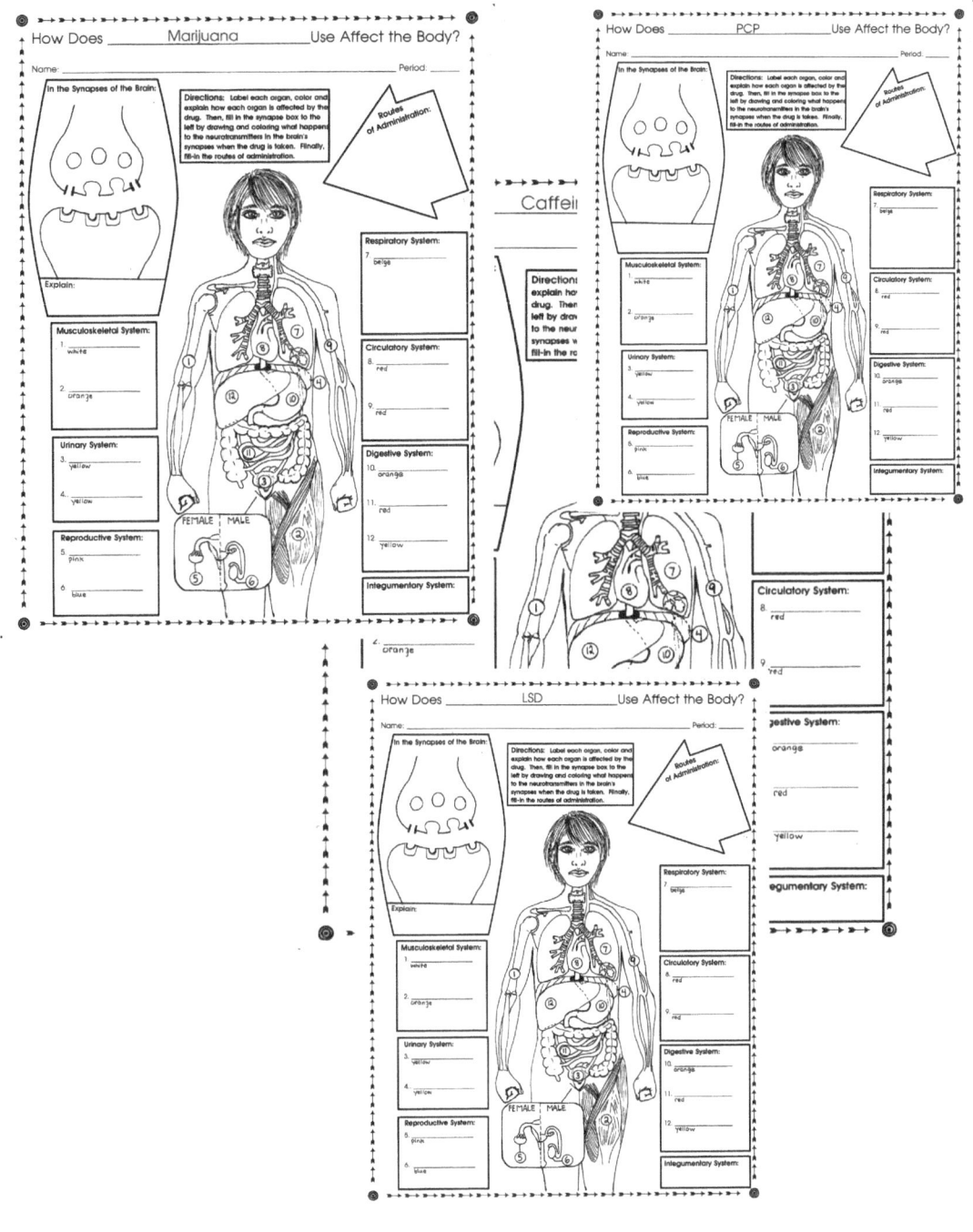

Depressants/Sedatives
(Alcohol, etc.)

Depressants are drugs that calm, relax and make the person sleepy. They tend to suppress central nervous system activity. Most of them are considered to be GABA agonists. In other words, they increase the inhibitory affects of GABA, our main inhibitory neurotransmitter that has a quieting effect on the brain. If you look at the GABA-gated chloride channel below, which is found on the membranes of post-synaptic neurons, you see that GABA, alcohol, barbiturates and benzodiazepines all have receptor sites on this protein. If any of these drugs are taken by a person, the binding of these molecules opens the chloride channel, allowing negatively charged Cl^- ions into the neuron's cell body. This change in charge pushes the neuron away from firing, and there is a quieting effect on the brain. That is why they are called GABA agonists (they help GABA do it's job). Some other side effects are drowsiness, falls, impaired coordination, suppressed breathing, nausea and vomiting. Depressants are often prescribed to treat both anxiety and insomnia. Repeated use of depressants will lead to physical dependence and possible psychological dependence. The abuse potential is pretty high.

GABA-gated Cl- channel

A- GABA in receptor site
B- Alcohol in receptor site
C- Benzodiazepine in site
D- Barbiturate in site
E- Cl- ion traveling
   through the channel
   protein in the neuron
   membrane

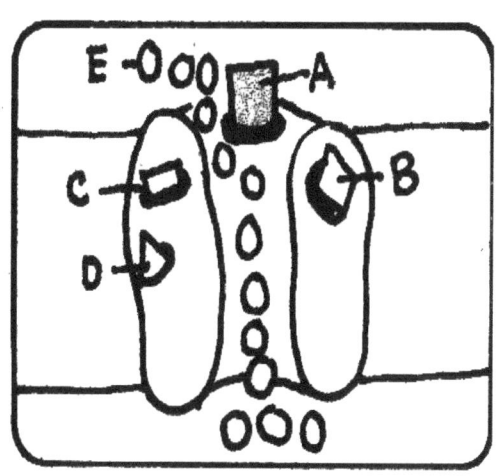

Perhaps the most popular and abused depressant is **alcohol.** Besides increasing the inhibitory affect of GABA, Alcohol binds to glutamate receptors on the post-synaptic membrane, preventing glutamate (an excitatory neurotransmitter) from attaching to the receptors and doing it's job to excite the post-synaptic neuron.

Alcohol at low doses, is associated with feelings of euphoria and happiness; but at higher doses, the person starts to feel sedated, a decrease in reaction time and visual acuity (bad for driving), and less alert.   There is less control over behavior. The person says and does things they wouldn't normally do! With excessive use, there might be loss of consciousness and difficulty remembering things. Of course, pregnant women who drink alcohol will run the risk of delivering a baby with Fetal Alcohol Syndrome.

As you start to fill-in the worksheet on **alcohol**, you will start to see the many body organs that are affected by this drug. Liver problems are major after using alcohol for a long time and cancer is possible in the bladder, stomach, and intestines. It leaves the bones and muscles weak, and can also weaken the heart. In very large amounts, alcohol can stop breathing and cause death!

 Be sure to show both ways that alcohol affects the synapse; it's affect on GABA, and it's affect on glutamate. They are both neurotransmitters, but one is inhibitory and one is excitatory.

Name: _____ Period: _____

In the Synapses of the Brain:

Explain:

Directions: Label each organ, color and explain how each organ is affected by the drug. Then, fill in the synapse box to the left by drawing and coloring what happens to the neurotransmitters in the brain's synapses when the drug is taken. Fiinally, fill-in the routes of administration.

Routes of Administration:

Musculoskeletal System:

1. _____
   white

2. _____
   orange

Urinary System:

3. _____
   yellow

4.. _____
   yellow

Reproductive System:

5. _____
   pink

6. _____
   blue

FEMALE   MALE

Respiratory System:

7. _____
   beige

Circulatory System:

8. _____
   red

9. _____
   red

Digestive System:

10. _____
    orange

11. _____
    red

12. _____
    yellow

Integumentary System:

Stimulants
(Amphetamines, Cocaine, Caffeine, Nicotine, Ecstasy, etc.)

Stimulants are drugs that tend to increase the overall levels of activity in a neuron. Many of them are dopamine agonists, which means that they increase the effects of dopamine, an excitatory neurotransmitter. Dopamine is a neurotransmitter often associated with pleasure, reward and craving, so they often are abused quite a bit. By different mechanisms, the end effect is more dopamine in the synapse to continuously stimulate the post-synaptic neuron. They might block the re-uptake pumps, push out extra dopamine, or do something similar to a few other neurotransmitters. They are all excitatory drugs in that they excite the next neuron to fire. We would consider drugs in this category central nervous system stimulants.

Some of the symptoms of stimulant abuse are euphoria, lots of energy, greater concentration, reduced appetite, able to stay awake, reward and craving, etc. But negative side-effects occur too, like anxiety, paranoia, hallucination, nausea, psychosis, high blood pressure, heart attack, confusion, agitation and possible death!

Amphetamines, including **methamphetamine**, are dopamine *agonists*. They invade the neurotransmitter's vesicles and push out extra dopamine through a re-uptake pump that has been reversed! In this way, more dopamine enters the synapse than usual and therefore, more dopamine effects like alertness, happiness and hunger are experienced. Body organs are affected by amphetamines too, as seen in the increased heart rate, increased blood pressure, acute liver failure, brittle bones and tense muscles, and urinary problems.

In recent years, **Methamphetamine** has become more popular. Illegal methamphetamine has a high potential for abuse since it is so addictive. High doses can cause psychosis, seizures, delusions and violent behavior. Meth can even cause breakdown of skeletal muscle and bleeding in the brain.

Caffeine is an adenosine *antagonist* in that it keeps adenosine (an inhibitory neurotransmitter in the brain) from attaching to it's receptors and thereby leaves us alert and awake. Besides keeping us alert and awake, it gives us energy, headache relief and suppresses appetite. It all sounds good, but the negative effects of **caffeine** are insomnia, anxiety, headaches upon withdrawal and a need to urinate a lot. Caffeine can also contribute to panic attacks and palpitations of the heart.

Cocaine is another dopamine *agonist* because it keeps dopamine in the synapse by blocking it from being pulled back into the vesicles after it has done it's job of stimulating receptors on the post-synaptic neuron. In other words, it blocks the re-uptake pumps. The effects of cocaine use are similar to amphetamines, such as euphoria, energy, strong emotions, contentment and all the negative side-effects like confusion, agitation, heart attack and liver failure. Besides these, the cartilage in the nose breaks down due to snorting **cocaine** and there can be shortness of breath. Skin ulcers can show up where the needle is injected and infertility of both sexes is known to happen. Blood vessels constrict, which causes high blood pressure and less blood supply to intestines causing nausea and diarrhea. Using a needle to inject cocaine brings the highest levels of dependence over smoking and snorting.

Ecstasy, also known as MDMA, makes serotonin (an excitatory neurotransmitter) re-uptake pumps work in reverse so that serotonin cannot be pulled out of the synapse and back into the vesicles. This causes overstimulation by serotonin and similar effects to the other stimulants with an increase in happiness, fullness, calmness and emotional well-being. Users claim that they have deep and unusual thoughts, inspiration, and contentment. Negative effects are the same as amphetamines plus brain damage and death due to hyperthermia and heart attack. Bones become brittle, muscles cramp and jaws clench. The scary thing about **ecstasy** is that the constant bombardment of the serotonin receptors causes them to retreat and retract into the cell membrane. This means that the person will not be as sensitive to the drug and will have to use more to get the same effect. The body thinks that increasing drug intake makes the few remaining receptor sites fire faster to restore the original reaction. We call this becoming tolerant to the drug, or *tolerance*. See the figure below. The one on the left is with early use of ecstasy and the one on the right shows that the receptors have retracted and more ecstasy is needed to try to restore the original reaction.

Nicotine (Tobacco) is an acetylcholine receptor agonist which means it increases the effects of acetylcholine.

Acetylcholine is a common neurotransmitter that normally decreases heart rate and increases memory and muscle contractions. We would say that acetylcholine is an excitatory neurotransmitter since it opens Na+ channels in the post-synaptic neuron, causing the nerve impulse to continue. Nicotine is also a dopamine (excitatory) *agonist* in that it causes more dopamine to be released when it attaches to nicotine receptors on the cell. **Nicotine** seems to act like both stimulant and relaxant. At first, it causes a release of *adrenaline* which causes the "fight or flight" syndrome, so the person feels alert and sharp. In it's promotion of dopamine effects, it increases the sensitivity of the brain's reward system, so the person feels pleasure. With increasing dosages, nicotine becomes a sedative and can decrease neuron activity.

Smoking cigarettes adds tar and many other chemicals to the mix, so the lungs of a smoker build up tar and can lead to lung cancer and emphysema. Since **nicotine** narrows the blood vessels (constriction), it's use can lead to heart attack and stroke. We also see an increase of ulcers in the digestive system and possible cancer in the bladder, cervix and testicles. Scientists say that bones become more brittle and muscles more weak due to less oxygen to the muscles from constricted blood vessels.

Name: _____ Period: _____

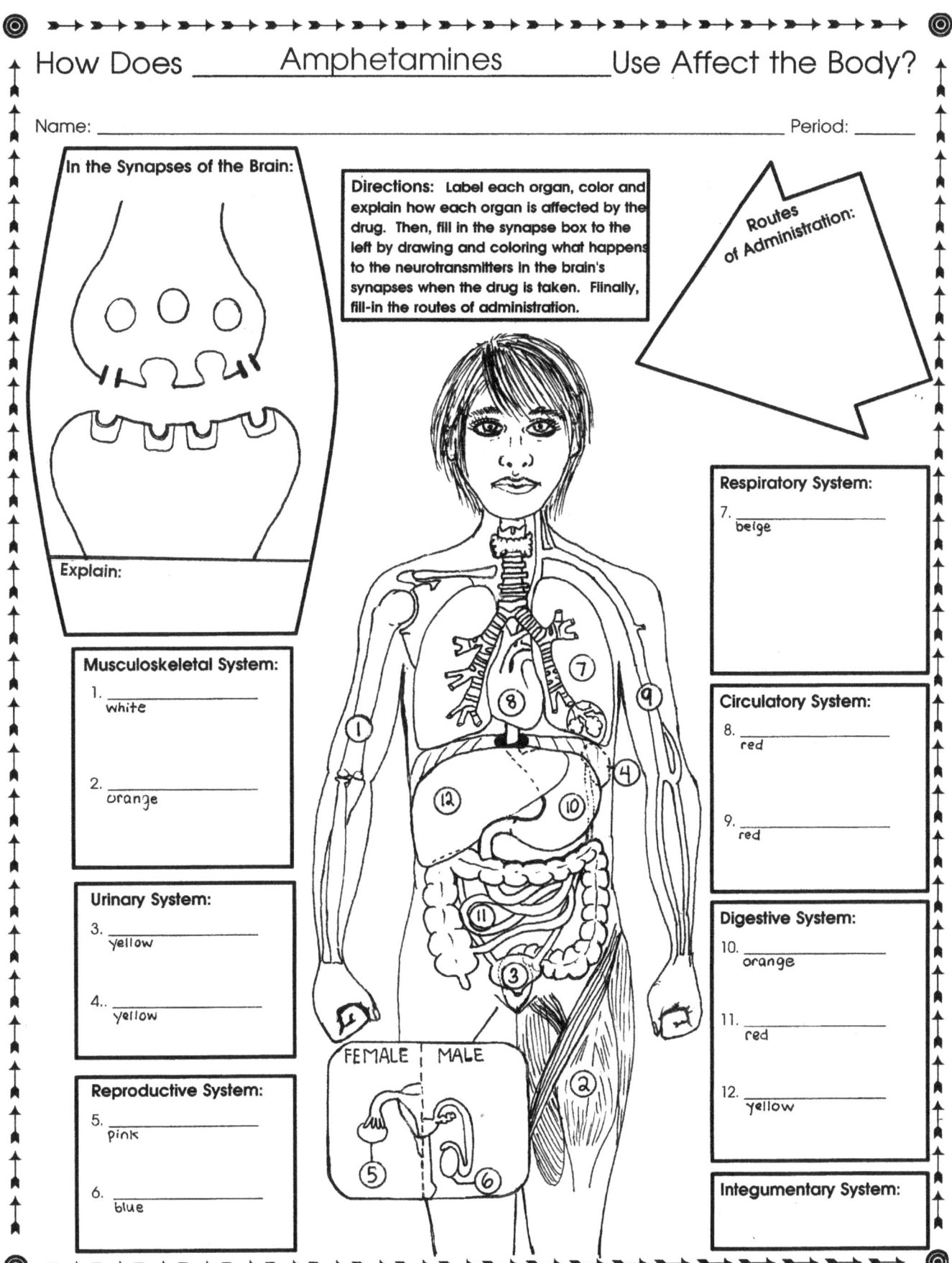

In the Synapses of the Brain:

Directions: Label each organ, color and explain how each organ is affected by the drug. Then, fill in the synapse box to the left by drawing and coloring what happens to the neurotransmitters in the brain's synapses when the drug is taken. Fiinally, fill-in the routes of administration.

Routes of Administration:

Explain:

Musculoskeletal System:
1. _____
   white

2. _____
   orange

Urinary System:
3. _____
   yellow

4.. _____
   yellow

Reproductive System:
5. _____
   pink

6. _____
   blue

FEMALE | MALE

Respiratory System:
7. _____
   beige

Circulatory System:
8. _____
   red

9. _____
   red

Digestive System:
10. _____
    orange

11. _____
    red

12. _____
    yellow

Integumentary System:

21

Name: _____ Period: _____

In the Synapses of the Brain:

Explain:

Directions: Label each organ, color and explain how each organ is affected by the drug. Then, fill in the synapse box to the left by drawing and coloring what happens to the neurotransmitters in the brain's synapses when the drug is taken. Fiinally, fill-in the routes of administration.

Routes of Administration:

Respiratory System:

7. _____
beige

Circulatory System:

8. _____
red

9. _____
red

Musculoskeletal System:

1. _____
white

2. _____
orange

Urinary System:

3. _____
yellow

4.. _____
yellow

Reproductive System:

5. _____
pink

6. _____
blue

FEMALE | MALE

Digestive System:

10. _____
orange

11. _____
red

12. _____
yellow

Integumentary System:

How Does _____Cocaine_____ Use Affect the Body?

Name: _____ Period: _____

In the Synapses of the Brain:

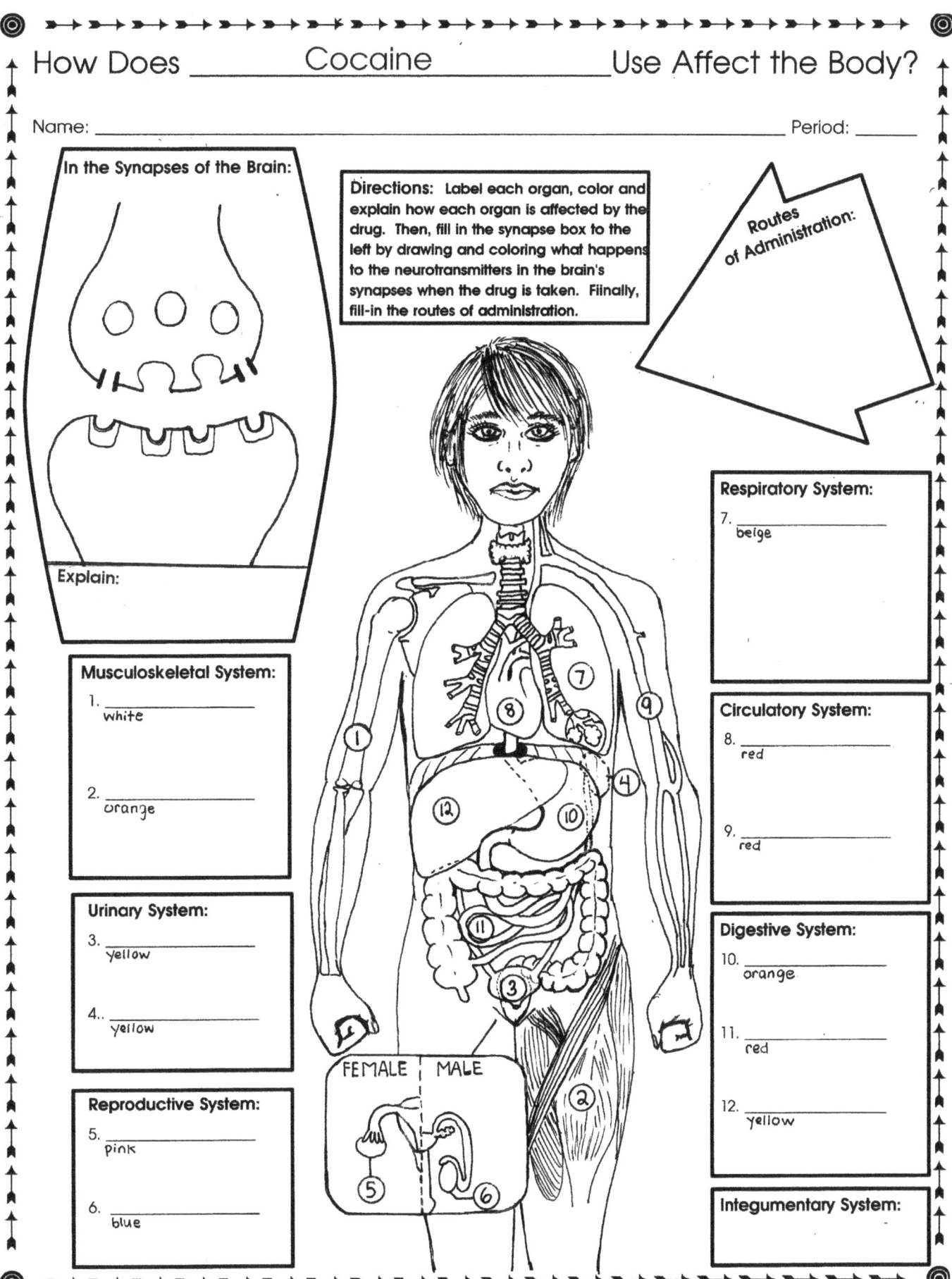

Explain:

Directions: Label each organ, color and explain how each organ is affected by the drug. Then, fill in the synapse box to the left by drawing and coloring what happens to the neurotransmitters in the brain's synapses when the drug is taken. Fiinally, fill-in the routes of administration.

Routes of Administration:

Musculoskeletal System:

1. _____
   white

2. _____
   orange

Urinary System:

3. _____
   yellow

4.. _____
   yellow

Reproductive System:

5. _____
   pink

6. _____
   blue

FEMALE | MALE

Respiratory System:

7. _____
   beige

Circulatory System:

8. _____
   red

9. _____
   red

Digestive System:

10. _____
   orange

11. _____
   red

12. _____
   yellow

Integumentary System:

Name: _____ Period: _____

In the Synapses of the Brain:

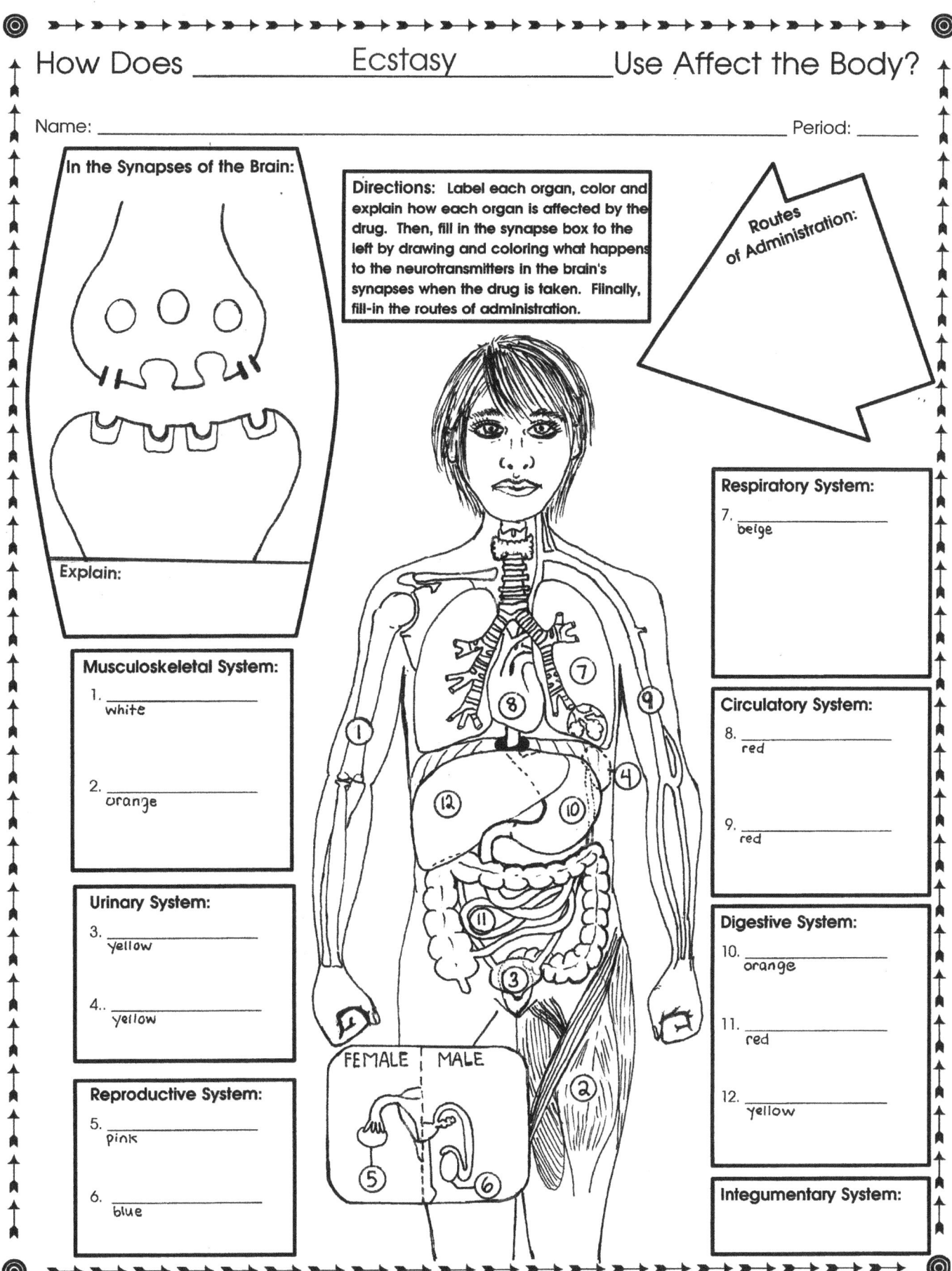

Directions: Label each organ, color and explain how each organ is affected by the drug. Then, fill in the synapse box to the left by drawing and coloring what happens to the neurotransmitters in the brain's synapses when the drug is taken. Finally, fill-in the routes of administration.

Routes of Administration:

Explain:

Musculoskeletal System:

1. _____
white

2. _____
orange

Urinary System:

3. _____
yellow

4.. _____
yellow

Reproductive System:

5. _____
pink

6. _____
blue

FEMALE | MALE

Respiratory System:

7. _____
beige

Circulatory System:

8. _____
red

9. _____
red

Digestive System:

10. _____
orange

11. _____
red

12. _____
yellow

Integumentary System:

How Does _____Tobacco_____ Use Affect the Body?

Name: _____ Period: _____

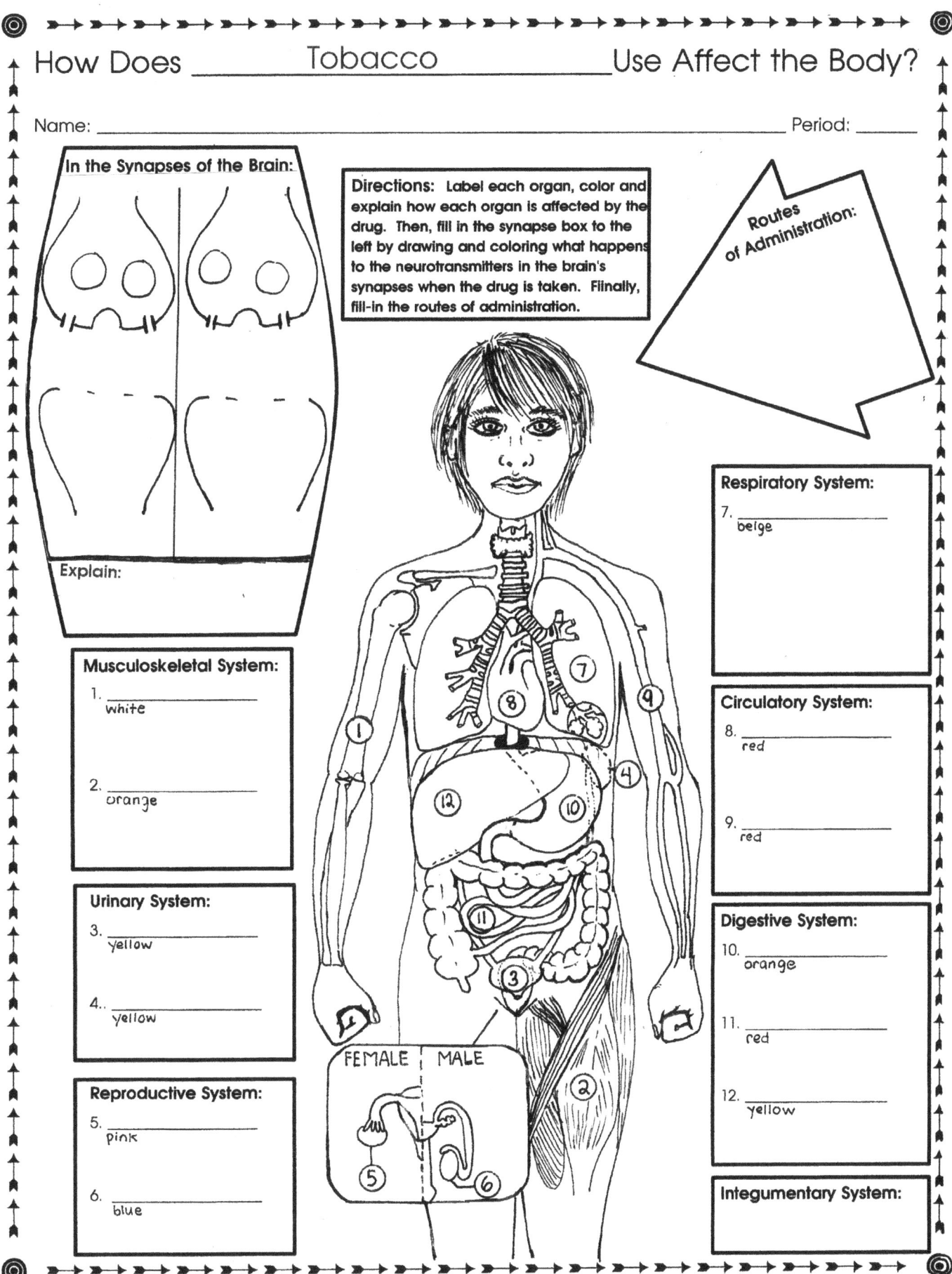

In the Synapses of the Brain:

Explain:

Directions: Label each organ, color and explain how each organ is affected by the drug. Then, fill in the synapse box to the left by drawing and coloring what happens to the neurotransmitters in the brain's synapses when the drug is taken. Finally, fill-in the routes of administration.

Routes of Administration:

Musculoskeletal System:
1. _____ white
2. _____ orange

Urinary System:
3. _____ yellow
4. _____ yellow

Reproductive System:
5. _____ pink
6. _____ blue

FEMALE | MALE

Respiratory System:
7. _____ beige

Circulatory System:
8. _____ red
9. _____ red

Digestive System:
10. _____ orange
11. _____ red
12. _____ yellow

Integumentary System:

Opioids/Narcotics
(Heroin, etc.)

Opioids also known as Narcotics include heroin (featured here), morphine, methadone, and codeine. They all have analgesic properties in that they decrease pain. They are opioid *antagonists* in that Opioid drugs like **heroin** mimic our body's own opioid chemicals that bind to opioid receptors reducing pain and producing euphoria. The natural opiates come from the poppy plant. Today, we have some synthetic versions of opiate drugs that are extremely abused like hydrocodone and oxycodone.

Heroin acts in two ways: it prevents pain by preventing Substance "P" from acting, and it bind to receptors on a GABA cell, preventing GABA release. When GABA isn't released, it can't inhibit dopamine release, and thus, dopamine gets released. We know what the effects of dopamine are: pleasure and reward. GABA is an inhibitory neurotransmitter because it prevents other neurotransmitters from firing. Look at the diagrams below and on the next page to understand the two ways in which **heroin** acts in the synapses of the brain.

Before heroin: GABA attaches to receptors on dopamine cell. No dopamine is released.

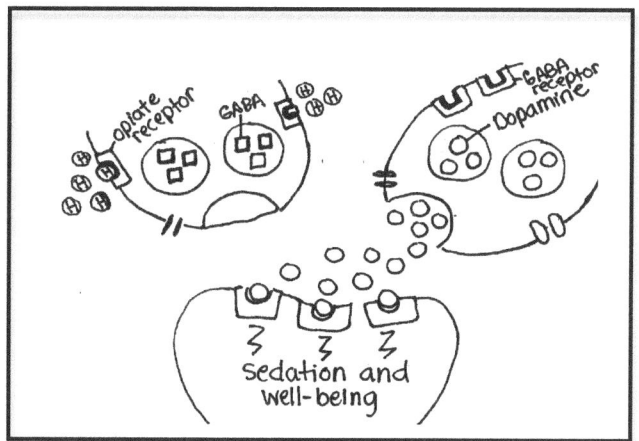

After heroin: heroin attaches to opiate receptors, GABA not released. GABA can't inhibit dopamine, so dopamine floods synapse = well-being

How Heroin Inhibits Pain

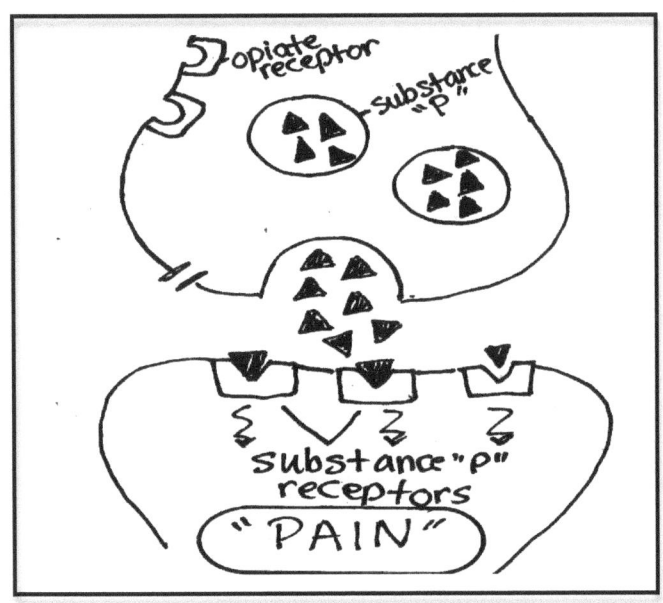

Substance P normally causes pain as it fits into it's receptors.

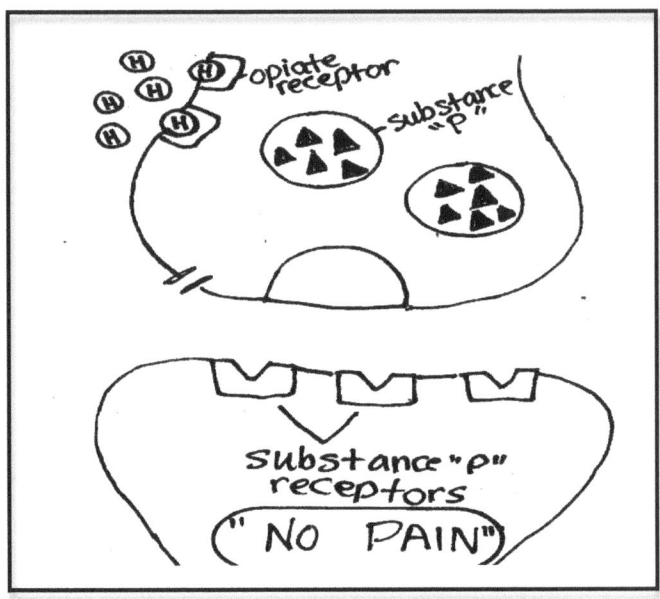

Heroin attaches to the opiate receptors on the substance P neuron and keep it from being released. The result is no pain.

Heroin causes euphoria, pain relief, calmness, relaxation, and sleepiness, but also causes nausea, constipation, vomiting, drowsiness and breathing suppression. If needles are used, infection of heart valves due to bacteria from dirty needles is possible and users have an increased risk of contracting tuberculosis and HIV. Veins can collapse at the site of injections. Users become constipated because their intestinal movements slow down. People can become so addicted to **heroin** that their entire life is devoted to getting and using Heroin.

Withdrawal from heroin is very difficult because it feels like having a bad flu. Some people are given a drug called Methadone, a synthetic opioid that is less euphorigenic than **heroin,** to help them withdraw from it.

How Does _____Heroin_____ Use Affect the Body?

Name: _____ Period: _____

In the Synapses of the Brain:

Explain:

Directions: Label each organ, color and explain how each organ is affected by the drug. Then, fill in the synapse box to the left by drawing and coloring what happens to the neurotransmitters in the brain's synapses when the drug is taken. Finally, fill-in the routes of administration.

Routes of Administration:

Musculoskeletal System:

1. _____
   white

2. _____
   orange

Urinary System:

3. _____
   yellow

4.. _____
   yellow

Reproductive System:

5. _____
   pink

6. _____
   blue

FEMALE | MALE

Respiratory System:

7. _____
   beige

Circulatory System:

8. _____
   red

9. _____
   red

Digestive System:

10. _____
    orange

11. _____
    red

12. _____
    yellow

Integumentary System:

Hallucinogens
(LSD, PCP, etc.)

Hallucinogens are drugs that result in intense alterations in sensory and perceptual experiences. Often, users experience vivid hallucinations and have a distorted perception of time. These drugs are varied in their action on the synapse and neurotransmitters. Some are serotonin *agonists* or *antagonists* while others are *antagonists* of the NMDA glutamate receptor. Serotonin is a excitatory neurotransmitter whose effects are an increase in happiness, fullness, and a lack of pain, while glutamate's effects are in the memory and learning areas. Glutamate is perhaps the most common excitatory neurotransmitter.

LSD (Lysergic acid diethylamine) mimics serotonin and sometimes acts like a serotonin *agonist* while at other times it can act like a serotonin *antagonist*. This seems confusing, but it is one reason why LSD has such complex sensory effects.

LSD mimics the shape of serotonin and binds to it's receptors on the post-synaptic neuron. There are a couple of different types of serotonin receptors in the brain and LSD interacts differently with them. It is mostly inhibitory in certain parts of the brain dealing with perception, but excitatory in other areas. On the worksheet sample, we see LSD mimicking serotonin, but in this example, acting like an antagonist (inhibitor). When it attaches to serotonin type 2 receptors, it inhibits them. It can be shown either way on the worksheet.

Users say that LSD causes feelings of novelty, inspiration, fast and disordered thinking, trances, perceptual anomalies like patterns moving, seeing sounds, smelling colors, etc. Users come up with crazy ideas and beliefs.

Negative effects like anxiety, insomnia, paranoia, flashbacks and temporary psychosis can occur too. Sometimes, users might feel nauseated and show tremors.

PCP (phencyclidine) is considered a "dissociative anesthetic" because it was originally used as an anesthetic, but was taken off the market because it caused dissociative hallucinations. PCP is considered to be a NMDA receptor antagonist which means it blocks the activity of the glutamate NMDA receptor. Glutamate normally attaches to the receptor and then Ca^{+2} ions are allowed to enter the post-synaptic neuron, causing the nerve to fire. But when PCP is in the synapse it attaches to the receptor right in the middle of the ion channel, blocking the ions from entering and thus stopping the impulse. The result is analgesic (pain reducing), dissociative effects and visual alterations. We say that this is "non-competitive" inhibition. See the diagrams below.

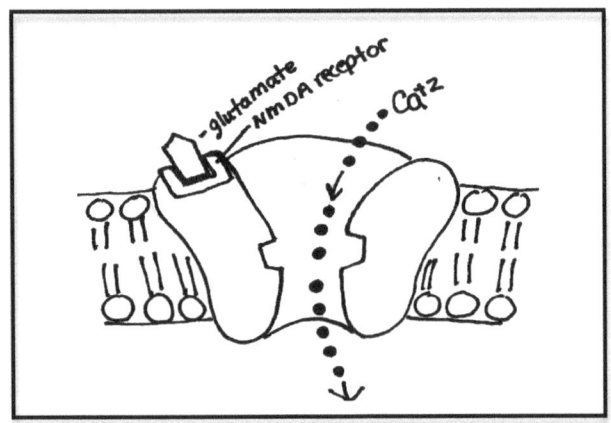

NMDA receptor complex: when glutamate binds to it's NMDA receptor, Calcium ions are allowed to enter, causing the nerve to fire

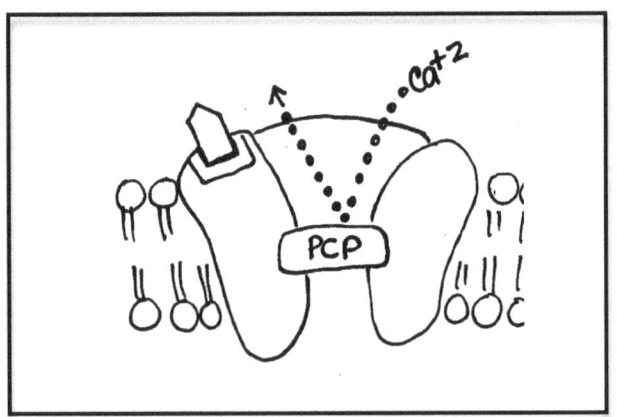

PCP attaches to it's receptor in the middle of the channel, preventing calcium flow

PCP causes the user to feel a distance from reality and the body. They also feel numb and have hallucinations. They may get nauseated, be violent, confused and have temporary psychosis. Some people go into a coma and get brain damage.

How Does _____LSD_____ Use Affect the Body?

Name: _____ Period: _____

In the Synapses of the Brain:

Explain:

Directions: Label each organ, color and explain how each organ is affected by the drug. Then, fill in the synapse box to the left by drawing and coloring what happens to the neurotransmitters in the brain's synapses when the drug is taken. Flinally, fill-in the routes of administration.

Routes of Administration:

Musculoskeletal System:
1. _____
   white

2. _____
   orange

Urinary System:
3. _____
   yellow

4.. _____
   yellow

Reproductive System:
5. _____
   pink

6. _____
   blue

FEMALE | MALE

Respiratory System:
7. _____
   belge

Circulatory System:
8. _____
   red

9. _____
   red

Digestive System:
10. _____
   orange

11. _____
   red

12. _____
   yellow

Integumentary System:

How Does _____PCP_____ Use Affect the Body?

Name: _____ Period: _____

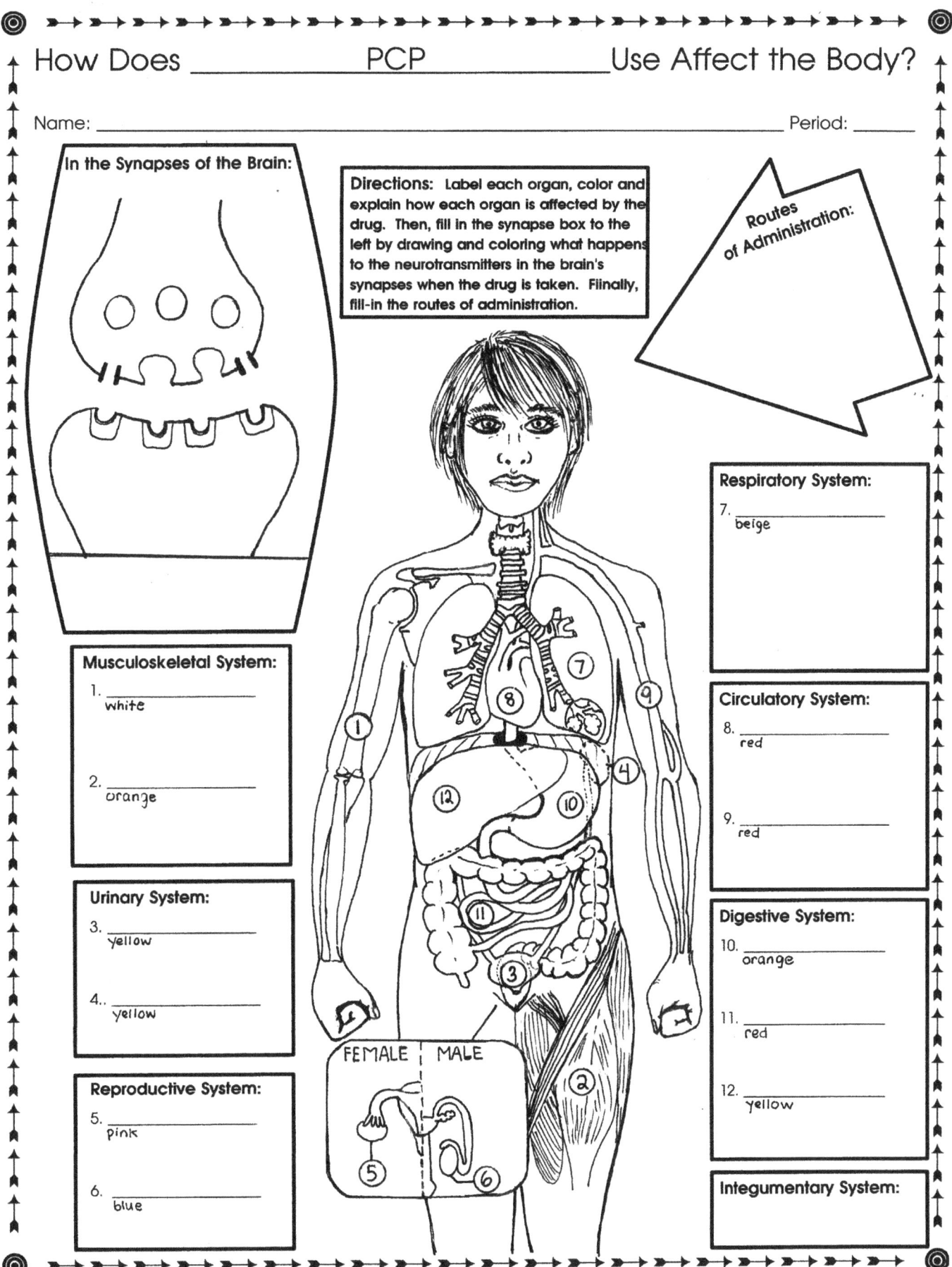

In the Synapses of the Brain:

Directions: Label each organ, color and explain how each organ is affected by the drug. Then, fill in the synapse box to the left by drawing and coloring what happens to the neurotransmitters in the brain's synapses when the drug is taken. Fiinally, fill-in the routes of administration.

Routes of Administration:

Musculoskeletal System:

1. _____
   white

2. _____
   orange

Urinary System:

3. _____
   yellow

4.. _____
   yellow

Reproductive System:

5. _____
   pink

6. _____
   blue

FEMALE MALE

Respiratory System:

7. _____
   beige

Circulatory System:

8. _____
   red

9. _____
   red

Digestive System:

10. _____
    orange

11. _____
    red

12. _____
    yellow

Integumentary System:

Cannabis (Marijuana)

Marijuana is a hard drug to put into a group because it sometimes acts as a depressant (slows down movement), other times as a stimulant (raises heart rate) and at times, a hallucinogen (can change perception). It is also affected by the strain of marijuana, the user's physiology and the amount taken. The active chemical in marijuana, called THC (tetrahydrocannabinol), is a GABA *antagonist* similar in action to heroin. THC binds to the cannabinoid receptors on the nerve cell (just like our body's natural cannabinoid called anandamide) that produces GABA. Remember that GABA is an inhibitory neurotransmitter and usually prevents dopamine from leaving the cell. When THC attaches to cannabinoid receptors, GABA cannot be released and therefore cannot keep dopamine from being released. Once dopamine is released, it moves across the synapse to it's own receptors and causes the nerve impulse to continue. The result is pleasure, relaxation, calmness and an enhanced appreciation of art. Of course, there can also be impairment of coordination, memory, thinking and reflexes.

Marijuana has other effects on the body too. The tar in marijuana damages the air sacs in the lungs and can cause emphysema and inflamed bronchial tubes. Reproductive organs can be affected too, such as lowering sperm count and altering the menstrual cycle. The heart rate increases and so does the blood pressure. There is some question as to whether it suppresses the immune system causing less resistance to illness.

Name: _____ Period: _____

In the Synapses of the Brain:

Explain:

Directions: Label each organ, color and explain how each organ is affected by the drug. Then, fill in the synapse box to the left by drawing and coloring what happens to the neurotransmitters in the brain's synapses when the drug is taken. Fiinally, fill-in the routes of administration.

Routes of Administration:

Musculoskeletal System:

1. _____
white

2. _____
orange

Urinary System:

3. _____
yellow

4.. _____
yellow

Reproductive System:

5. _____
pink

6. _____
blue

FEMALE | MALE

Respiratory System:

7. _____
beige

Circulatory System:

8. _____
red

9. _____
red

Digestive System:

10. _____
orange

11. _____
red

12. _____
yellow

Integumentary System:

Sample Answer Sheets

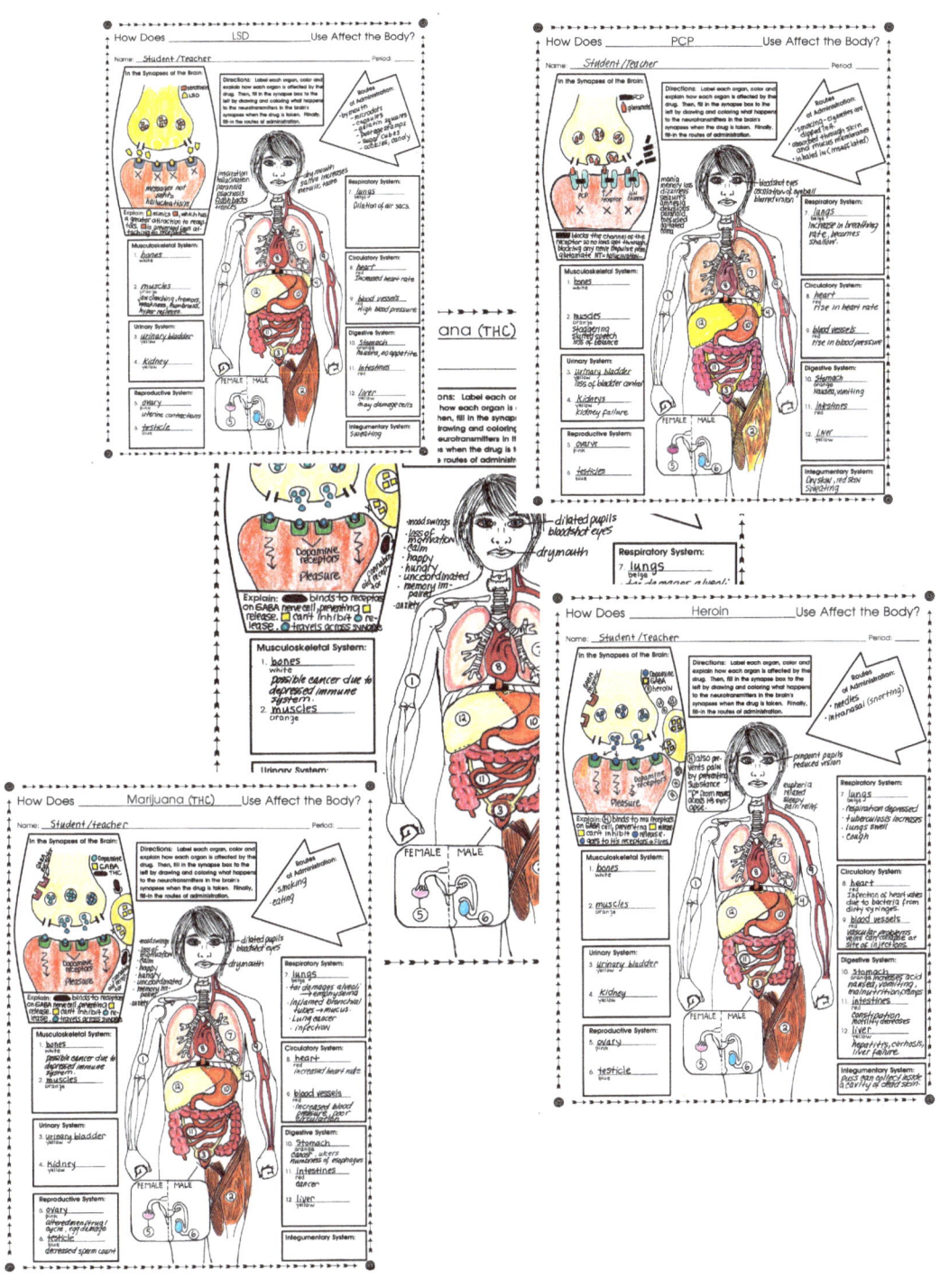

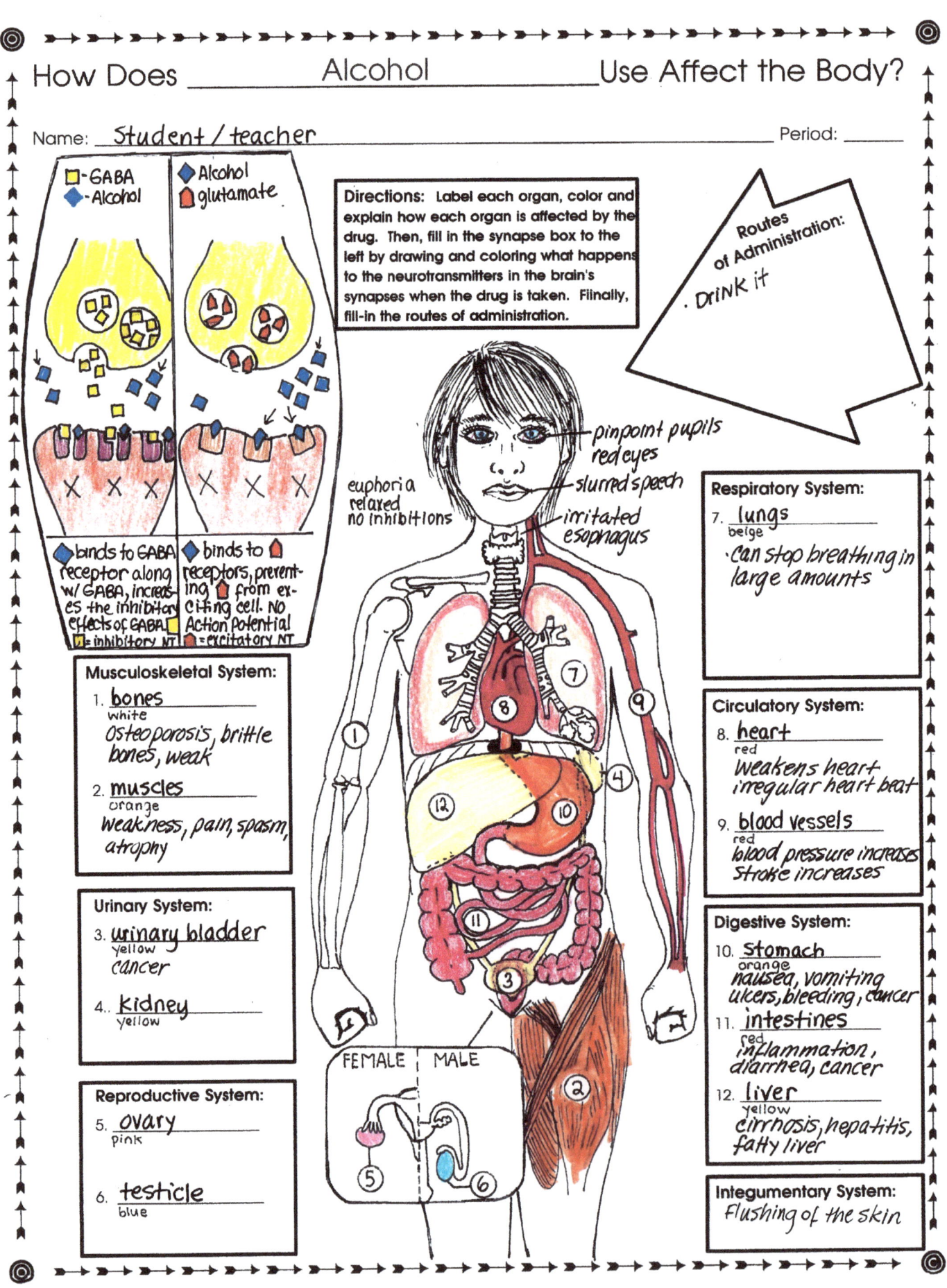

How Does _____Alcohol_____ Use Affect the Body?

Name: Student / teacher   Period: _____

- ▢ - GABA
- ◆ - Alcohol

- ◆ Alcohol
- ▲ glutamate

Directions: Label each organ, color and explain how each organ is affected by the drug. Then, fill in the synapse box to the left by drawing and coloring what happens to the neurotransmitters in the brain's synapses when the drug is taken. Finally, fill-in the routes of administration.

Routes of Administration:
- Drink it

◆ binds to GABA receptor along w/ GABA, increases the inhibitory effects of GABA ▢
▢ = inhibitory NT

◆ binds to ▲ receptors, preventing ▲ from exciting cell. No Action potential
▲ = excitatory NT

euphoria
relaxed
no inhibitions

pinpoint pupils
red eyes
slurred speech
irritated esophagus

Musculoskeletal System:

1. **bones**
   white
   Osteoporosis, brittle bones, weak

2. **muscles**
   orange
   weakness, pain, spasm, atrophy

Urinary System:

3. **urinary bladder**
   yellow
   cancer

4. **kidney**
   yellow

Reproductive System:

5. **ovary**
   pink

6. **testicle**
   blue

FEMALE | MALE

Respiratory System:

7. **lungs**
   beige
   · Can stop breathing in large amounts

Circulatory System:

8. **heart**
   red
   weakens heart
   irregular heart beat

9. **blood vessels**
   red
   blood pressure increases
   Stroke increases

Digestive System:

10. **stomach**
    orange
    nausea, vomiting
    ulcers, bleeding, cancer

11. **intestines**
    red
    inflammation, diarrhea, cancer

12. **liver**
    yellow
    cirrhosis, hepatitis, fatty liver

Integumentary System:
Flushing of the skin

(and methamphetamines)

Name: Student / Teacher _____ Period: _____

In the Synapses of the Brain:

● amphetamine
○ dopamine

receptors

pleasure

Explain: ● invade vesicles, push out extra dopamine out through reversed pump. Excess dopamine at receptors causes increase in pleasure. ● also blocks pump.

Directions: Label each organ, color and explain how each organ is affected by the drug. Then, fill in the synapse box to the left by drawing and coloring what happens to the neurotransmitters in the brain's synapses when the drug is taken. Fiinally, fill-in the routes of administration.

Routes of Administration:
1. swallow pills
2. snort (methamphetamine)
3. inject
4. smoke (methamphetamine)

violence
seizures
psychosis
brain damage
energetic
euphoria
anxiety
paranoia
stroke

bloodshot eyes, dilated pupils
acne
dry mouth
grinding teeth (bruxism)

Musculoskeletal System:

1. _bones_
white
become brittle, prone to breaking.

2. _muscles_
orange
tension, twitching, numbness, tremors
tightness of jaw muscles

Urinary System:

3. _urinary bladder_
yellow
infection
incontinance (can't hold urine
4. _kidney_
yellow

Reproductive System:

5. _ovary_
pink
elevates sexual arousal

6. _testicle_
blue
elevates sexual arousal

FEMALE | MALE

Respiratory System:

7. _lungs_
beige
rapid breathing
pulmonary edema (fluid in lungs)

Circulatory System:

8. _heart_
red
increased heart rate (tachycardia)

9. _blood vessels_
red
increase in blood pressure
constriction and strain of blood vessels

Digestive System:

10. _Stomach_
orange

11. _intestines_
red
diarrhea or constipation

12. _liver_
yellow
acute liver failure from IV drug use.

Integumentary System:
pimples, dry, itchy, pallor, feeling of bugs under skin.

How Does _____ Caffeine _____ Use Affect the Body?

Name: _____ Period: _____

48

In the Synapses of the Brain:

■ - caffeine
▲ - Adenosine

Adenosine normally makes us tired, sleepy.

Explain: Caffeine keeps Adenosine from attaching to its receptors since it fits into receptors → we are alert, awake

Directions: Label each organ, color and explain how each organ is affected by the drug. Then, fill in the synapse box to the left by drawing and coloring what happens to the neurotransmitters in the brain's synapses when the drug is taken. Finally, fill-in the routes of administration.

Routes of Administration:
· By mouth, in coffee, tea, chocolate, cola, other soda, energy drinks.
· In lotions to reduce itching.

insomnia
anxiety
headaches
alert
energetic

stained teeth
harms voice quality

Musculoskeletal System:
1. bones
   white

2. muscles
   orange

Urinary System:
3. urinary bladder
   yellow
   increases urine flow

4. kidney
   yellow

Reproductive System:
5. ovary
   pink

6. testicle
   blue

FEMALE | MALE

Respiratory System:
7. lung
   beige
   may alter breathing rate.

Circulatory System:
8. heart
   red
   Since caffeine raises epinephrine (adrenalin) it stimulates heart rate, can cause palpitations.

9. blood vessels
   red
   increases blood pressure.

Digestive System:
10. stomach
    orange
    Activates acid production → heartburn

11. intestines
    red

12. liver
    yellow

Integumentary System:
Possible flushing

How Does ___Cocaine___ Use Affect the Body?

Name: __Student/teacher_____ Period: _____

In the Synapses of the Brain:

🔴 cocaine
🔵 dopamine

vesicle

trans-
mitting
neuron

Synapse

Dopamine receptors

pleasure

receiving neuron

Explain: Cocaine blocks opening to re-uptake pumps so that dopamine keeps stimulating it's receptors → pleasure

Directions: Label each organ, color and explain how each organ is affected by the drug. Then, fill in the synapse box to the left by drawing and coloring what happens to the neurotransmitters in the brain's synapses when the drug is taken. Flinally, fill-in the routes of administration.

Routes of Administration:
1. smoking -43%
2. snorting -32%
3. needle-25%-highest levels of dependence

dilated pupils

cartilage in nose breaks down.

hoarseness, sore throat, bruxism of teeth

agitation confusion energetic euphoria

Musculoskeletal System:

1. __bones__
   white
   bone marrow erythro-poiesis (producing RBCs)
2. __muscles__
   orange
   lethargy, tired mus-cles.

Urinary System:

3. __urinary bladder__
   yellow
   increased need to urinate
4. __kidney__
   yellow

Reproductive System:

5. __ovary__
   pink
   infertility

6. __testicle__
   blue
   risk of impotence, infertility

FEMALE | MALE

Respiratory System:

7. __lung__
   beige
   Cough, shortness of breath (dyspnea). Damage to lung cells. Asthma, broncho-spasm, chest pain.

Circulatory System:

8. __heart__
   red
   Increased heart rate, blood pressure → heart attack, failure.
9. __blood vessels__
   red
   constriction of vessels, increasing blood pressure

Digestive System:

10. __stomach__
    orange
    hunger

11. __intestines__
    red
    Blood supply is reduced → nausea, diar-rhea.
12. __liver__
    yellow
    can lead to cirrhosis, hepatitis, cancer of liver.

Integumentary System:
trackmarks, skin ulcers where needle injects, feels like bugs under skin

Name: _Student/Teacher_____ Period: _____

In the Synapses of the Brain:

☑ ecstasy
■ serotonin

Directions: Label each organ, color and explain how each organ is affected by the drug. Then, fill in the synapse box to the left by drawing and coloring what happens to the neurotransmitters in the brain's synapses when the drug is taken. Fiinally, fill-in the routes of administration.

Routes of Administration:
· mouth (pills)

→ dilated pupils

→ teeth grinding

euphoria
energetic
emotional
confusion
agitation
brain damage

Explain: ☑ makes serotonin pumps work in reverse and don't allow re-uptake of ■, so ■ stays in synapse, stimulating → euphoria

Respiratory System:
7. _lungs_____
   beige
   may damage cells.

Musculoskeletal System:
1. _bones_____
   white
   calcification, brittle bones

2. _muscles_____
   orange
   tight muscles, tremors, cramping jaw clenching

Circulatory System:
8. _heart_____
   red
   increased heart rate possible heart attack

9. _blood vessels___
   red
   increases blood pressure

Urinary System:
3. _urinary bladder_
   yellow
   urinary retention

4. _Kidney_____
   yellow

Digestive System:
10. _Stomach_____
    orange
    Nausea, vomiting, loss of appetite

11. _Intestines_____
    red

12. _liver_____
    yellow

Reproductive System:
5. _ovary_____
   pink
   can affect menstruation

6. _testicle_____
   blue

FEMALE   MALE

Integumentary System:
sweating

50

Name: Student/Teacher _____ Period: _____

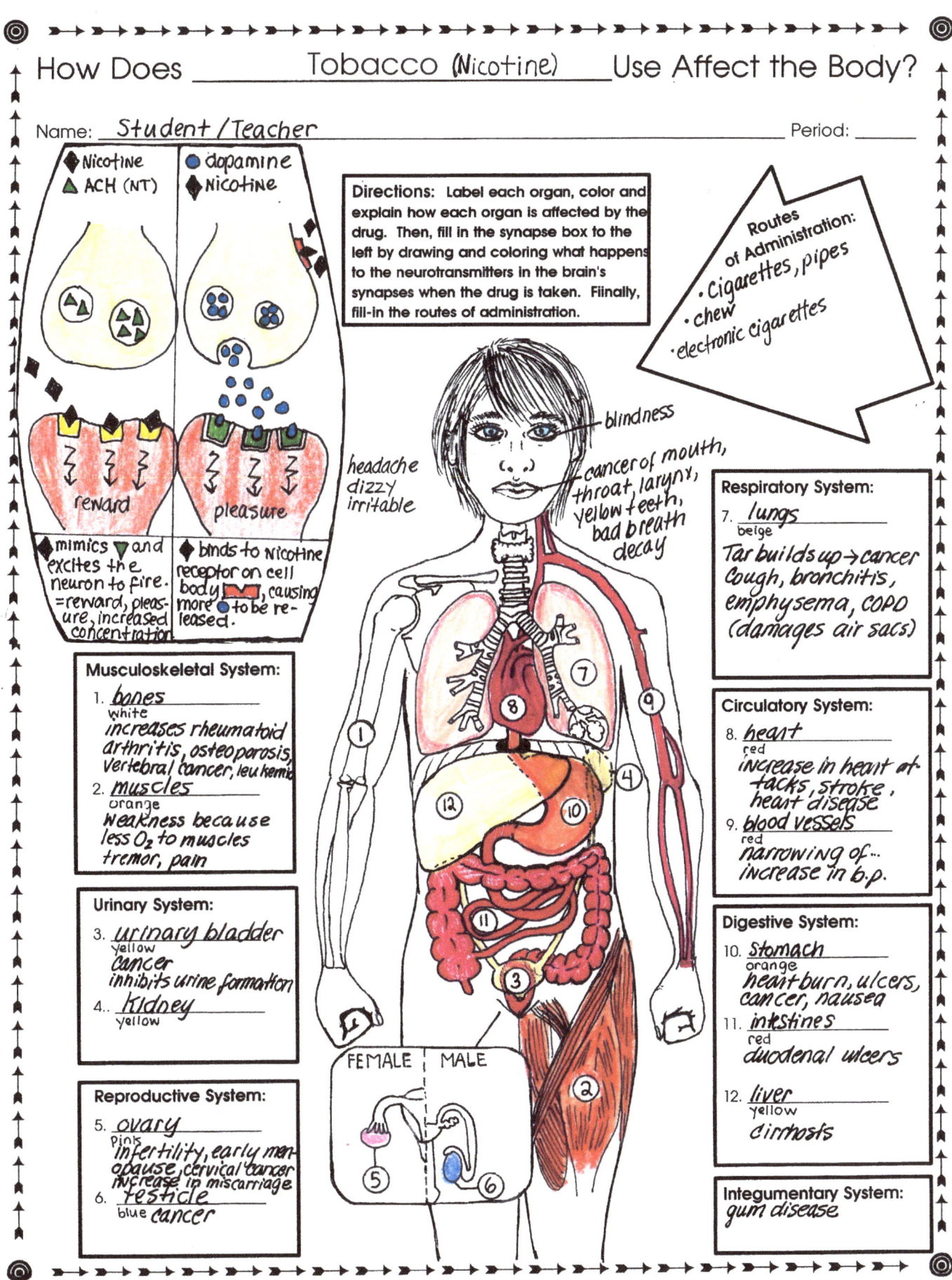

◆ Nicotine
▲ ACH (NT)

● dopamine
◆ Nicotine

Directions: Label each organ, color and explain how each organ is affected by the drug. Then, fill in the synapse box to the left by drawing and coloring what happens to the neurotransmitters in the brain's synapses when the drug is taken. Fiinally, fill-in the routes of administration.

Routes of Administration:
• Cigarettes, pipes
• chew
• electronic cigarettes

reward   pleasure

◆ mimics ▼ and excites the neuron to fire. = reward, pleasure, increased concentration

◆ binds to Nicotine receptor on cell body 🔴, causing more ● to be released.

headache
dizzy
irritable

blindness

cancer of mouth, throat, larynx, yellow teeth, bad breath decay

Respiratory System:
7. _lungs_
beige
Tar builds up → cancer
Cough, bronchitis, emphysema, COPD (damages air sacs)

Musculoskeletal System:
1. _bones_
white
increases rheumatoid arthritis, osteoporosis, vertebral cancer, leukemia
2. _muscles_
orange
weakness because less O_2 to muscles tremor, pain

Circulatory System:
8. _heart_
red
increase in heart attacks, stroke, heart disease
9. _blood vessels_
red
narrowing of... increase in b.p.

Urinary System:
3. _urinary bladder_
yellow
cancer
inhibits urine formation
4. _kidney_
yellow

Digestive System:
10. _stomach_
orange
heartburn, ulcers, cancer, nausea
11. _intestines_
red
duodenal ulcers
12. _liver_
yellow
cirrhosis

FEMALE | MALE

Reproductive System:
5. _ovary_
pink
infertility, early menopause, cervical cancer increase in miscarriage
6. _testicle_
blue cancer

Integumentary System:
gum disease

Name: Student/Teacher _____ Period: _____

In the Synapses of the Brain:

- Dopamine
- GABA
- heroIN

GABA receptor

Dopamine receptors

Pleasure

Explain: (H) binds to mu receptors on GABA cell, preventing ☐ release.
- ☐ can't inhibit ● release.
- ● goes to it's receptors → Fires

Directions: Label each organ, color and explain how each organ is affected by the drug. Then, fill in the synapse box to the left by drawing and coloring what happens to the neurotransmitters in the brain's synapses when the drug is taken. Fiinally, fill-in the routes of administration.

(H) also prevents pain by preventing Substance "P" from moving across it's synapse.

Routes of Administration:
- needles
- intranasal (snorting)

pinpoint pupils reduced vision

euphoria
relaxed
sleepy
pain relief

Respiratory System:
7. lungs
   beige
- respiration depressed
- tuberculosis increases
- lungs swell
- cough

Circulatory System:
8. heart
   red
   Infection of heart valves due to bacteria from dirty syringes.
9. blood vessels
   red
   vascular problems veins can collapse at site of injections.

Digestive System:
10. stomach
    orange Increases acid
    nausea, vomiting, malnutrition, cramps
11. intestines
    red
    constipation motility decreases
12. liver
    yellow
    hepatitis, cirrhosis, liver failure

Integumentary System:
puss can collect inside a cavity of dead skin.

Musculoskeletal System:
1. bones
   white

2. muscles
   orange

Urinary System:
3. urinary bladder
   yellow

4. kidney
   yellow

Reproductive System:
5. ovary
   pink

6. testicle
   blue

FEMALE | MALE

How Does _____LSD_____ Use Affect the Body?

Name: __Student/Teacher_____ Period: _____

In the Synapses of the Brain:

🟧 serotonin
🔼 LSD

messages not
sent =
hallucinations

Explain: 🔼 mimics 🟧, which has a greater attraction to receptors. 🟧 is prevented from attaching to receptors.

Directions: Label each organ, color and explain how each organ is affected by the drug. Then, fill in the synapse box to the left by drawing and coloring what happens to the neurotransmitters in the brain's synapses when the drug is taken. Finally, fill-in the routes of administration.

Routes of Administration:
• by mouth
  - microdots
  - capsules
  - gelatin squares
  - postage stamps
  - sugar cubes
  - cookies, candy

inspiration
hallucination
paranoia
psychosis
flash backs
trances

- dry mouth
saliva increases
metallic taste

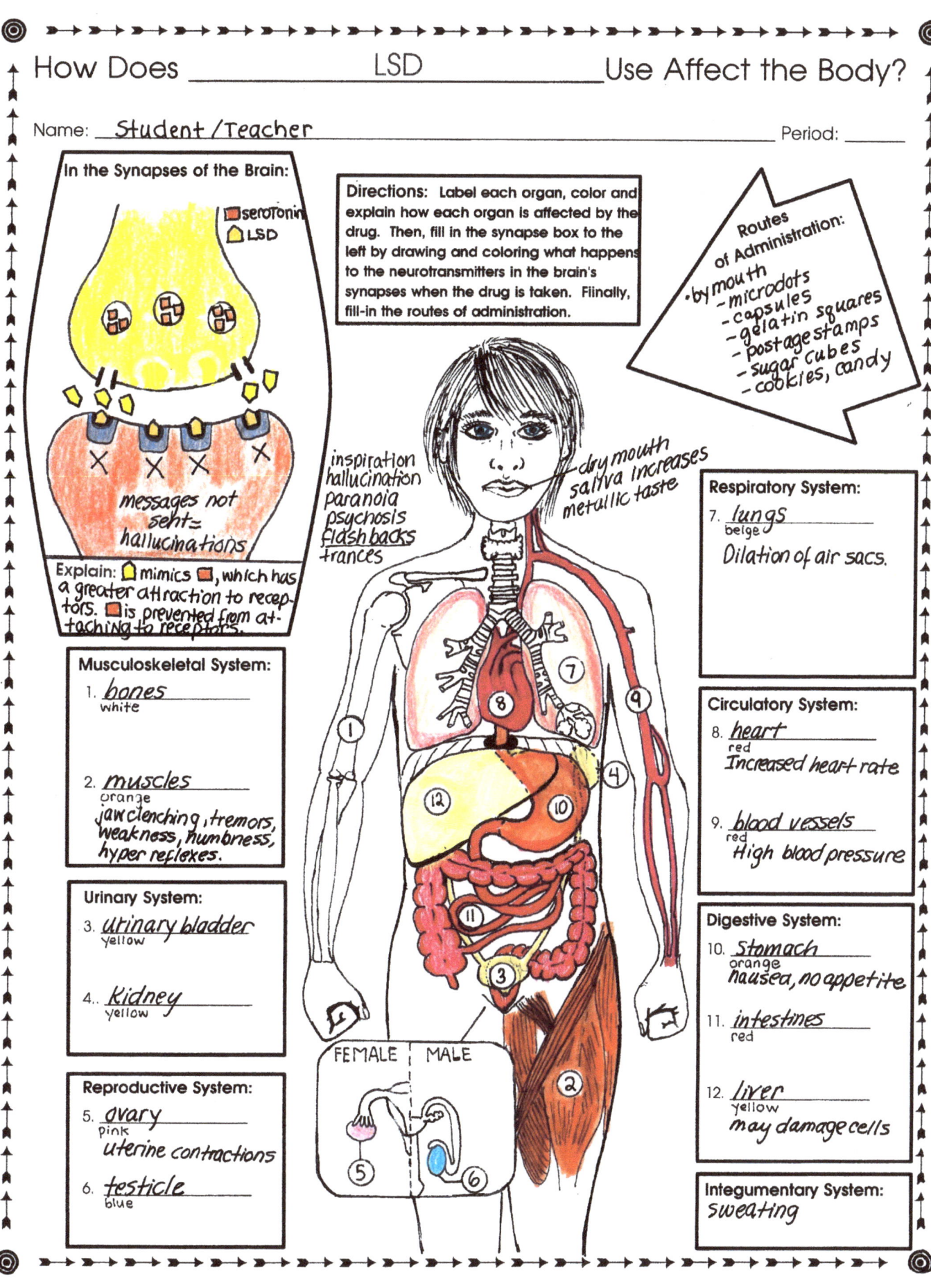

Musculoskeletal System:

1. _bones_
   white

2. _muscles_
   orange
   jaw clenching, tremors, weakness, numbness, hyper reflexes.

Urinary System:

3. _urinary bladder_
   yellow

4. _kidney_
   yellow

Reproductive System:

5. _ovary_
   pink
   uterine contractions

6. _testicle_
   blue

FEMALE | MALE

Respiratory System:

7. _lungs_
   beige
   Dilation of air sacs.

Circulatory System:

8. _heart_
   red
   Increased heart rate

9. _blood vessels_
   red
   High blood pressure

Digestive System:

10. _stomach_
    orange
    nausea, no appetite

11. _intestines_
    red

12. _liver_
    yellow
    may damage cells

Integumentary System:
sweating

How Does _____PCP_____ Use Affect the Body?

Name: __Student/Teacher_____ Period: _____

In the Synapses of the Brain:

■ PCP
🔺 glutamate

PCP   receptor   ion channel
 X   X   X

■ blocks the channel of the receptor so no ions get through, blocking any nerve impulse from glutamate NT = hallucination...

Directions: Label each organ, color and explain how each organ is affected by the drug. Then, fill in the synapse box to the left by drawing and coloring what happens to the neurotransmitters in the brain's synapses when the drug is taken. Fiinally, fill-in the routes of administration.

Routes of Administration:
• smoking - cigarettes are dipped in it.
• absorbed through skin and mucus membranes
• inhaled in (insufflated)

mania
memory loss
dizziness
seizures
amnesia
delusions
paranoid
confused
agitated
coma

bloodshot eyes
oscillation of eyeball
blurred vision

Musculoskeletal System:

1. __bones__
white

2. __muscles__
orange
staggering
slurred speech
loss of balance

Urinary System:

3. __urinary bladder__
yellow
loss of bladder control

4. __kidneys__
yellow
kidney failure

Reproductive System:

5. __ovarys__
pink

6. __testicles__
blue

FEMALE | MALE

Respiratory System:

7. __lungs__
beige
increase in breathing rate, becomes shallow.

Circulatory System:

8. __heart__
red
rise in heart rate

9. __blood vessels__
red
rise in blood pressure

Digestive System:

10. __stomach__
orange
Nausea, vomiting

11. __intestines__
red

12. __liver__
yellow

Integumentary System:
Dry skin, red skin
sweating

Name: _Student/teacher_____ Period: _____

In the Synapses of the Brain:

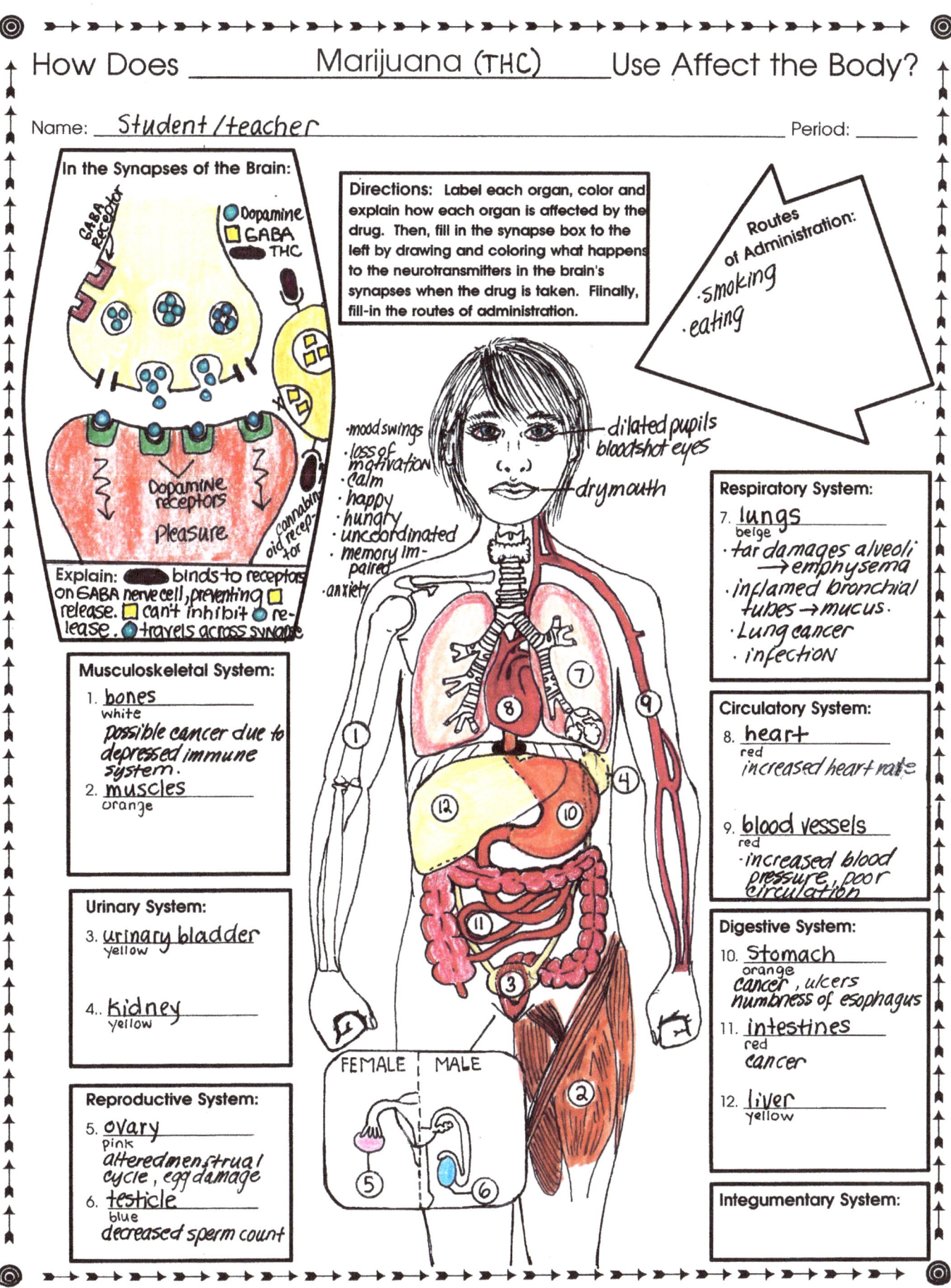

- Dopamine
- GABA
- THC

GABA Receptor

Dopamine receptors
Pleasure

Cannabinoid receptor

Explain: ⬛ binds to receptors on GABA nerve cell, preventing 🟨 release. 🟨 can't inhibit 🔵 release. 🔵 travels across synapse

Directions: Label each organ, color and explain how each organ is affected by the drug. Then, fill in the synapse box to the left by drawing and coloring what happens to the neurotransmitters in the brain's synapses when the drug is taken. Fiinally, fill-in the routes of administration.

Routes of Administration:
· smoking
· eating

· mood swings
· loss of motivation
· calm
· happy
· hungry
· uncoordinated
· memory impaired
· anxiety

dilated pupils
bloodshot eyes

dry mouth

Respiratory System:
7. __lungs__
   beige
· tar damages alveoli
  → emphysema
· inflamed bronchial tubes → mucus.
· Lung cancer
· infection

Circulatory System:
8. __heart__
   red
   increased heart rate

9. __blood vessels__
   red
· increased blood pressure, poor circulation

Digestive System:
10. __Stomach__
    orange
    cancer, ulcers
    numbness of esophagus
11. __intestines__
    red
    cancer
12. __liver__
    yellow

Musculoskeletal System:
1. __bones__
   white
   possible cancer due to depressed immune system.
2. __muscles__
   orange

Urinary System:
3. __urinary bladder__
   yellow
4. __kidney__
   yellow

Reproductive System:
5. __ovary__
   pink
   altered menstrual cycle, egg damage
6. __testicle__
   blue
   decreased sperm count

FEMALE | MALE